Dear parents,

As a mom and as an educator, I am ve
Workbook series with all of you. I developed this series for my two kids in elementary school, utilizing all of my knowledge and experience that I have gained while studying and working in the fields of Elementary Education and Gifted Education in South Korea as well as in the United States.

While raising my kids in the U.S., I had great disappointment and dissatisfaction about the math curriculum in the public schools. Based on my analysis, students cannot succeed in math with the current school curriculum because there is no sequential building up of fundamental skills. This is akin to building a castle on sand. So instead, I wanted to find a good workbook, but couldn't. And I also tried to find a tutor, but the price was too expensive for me. These are the reasons why I decided to make the Tiger Math series on my own.

The Tiger Math series was designed based on my three beliefs toward elementary math education.

1. It is extremely important to build foundation of math by acquiring a sense of numbers and mastering the four operation skills in terms of addition, subtraction, multiplication, and division.
2. In math, one should go through all steps in order, step by step, and cannot jump from level 1 to 3.
3. Practice math every day, even if only for 10 minutes.

If you feel that you don't know where your child should start, just choose a book in the Tiger Math series where your child thinks he/she can complete most of the material. And encourage your child to do only 2 sheets every day. When your child finishes the 2 sheets, review them together and encourage your child about his/her daily accomplishment.

I hope that the Tiger Math series can become a stepping stone for your child in gaining confidence and for making them interested in math as it has for my kids. Good luck!

Michelle Y. You, Ph.D.
Founder and CEO of Tiger Math

ACT scores show that only one out of four high school graduates are prepared to learn in college. This preparation needs to start early. In terms of basic math skills, being proficient in basic calculation means a lot. Help your child succeed by imparting basic math skills through hard work.

Sungwon S. Kim, Ph.D.
Engineering professor

# Level F – 1:  Plan of Study

| | Goal A | Practice division with 2 digit divisors. (Week 1 ~ 2) |
|---|---|---|
| | Goal B | Practice mixed operations with +, −, ×, ÷, and brackets with natural numbers. (Week 3 ~ 4) |

| Week 1 | Day | Tiger Session | | Topic | Goals |
|---|---|---|---|---|---|
| | Mon | 1 | 2 | 2 digits ÷ 2 digits | Divisors: 21, 22  (1 digit quotient) |
| | Tue | 3 | 4 | 3 digits ÷ 2 digits | Divisors: 31, 32  (1 digit quotient) |
| | Wed | 5 | 6 | 3 digits ÷ 2 digits | Divisors: 11 ~ 49  (1 digit quotient) |
| | Thu | 7 | 8 | 3 digits ÷ 2 digits | Divisors: 50 ~ 99  (1 digit quotient) |
| | Fri | 9 | 10 | 3 digits ÷ 2 digits | Divisors: 11 ~ 99  (1 digit quotient) |

| Week 2 | Day | Tiger Session | | Topic | Goal |
|---|---|---|---|---|---|
| | Mon | 11 | 12 | 4 digits ÷ 2 digits | Divisors: 21, 22  (2 digit quotient) |
| | Tue | 13 | 14 | 4 digits ÷ 2 digits | Divisors: 31, 32  (2 digit quotient) |
| | Wed | 15 | 16 | 4 digits ÷ 2 digits | Divisors: 11 ~ 49  (2 digit quotient) |
| | Thu | 17 | 18 | 4 digits ÷ 2 digits | Divisors: 50 ~ 99  (2 digit quotient) |
| | Fri | 19 | 20 | 4 digits ÷ 2 digits | Divisors: 11 ~ 99  (2 digit quotient) |

| Week 3 | Day | Tiger Session | | Topic | Goal |
|---|---|---|---|---|---|
| | Mon | 21 | 22 | Mixed operation | +, −, ( ) |
| | Tue | 23 | 24 | Mixed operation | ×, ÷, ( ) |
| | Wed | 25 | 26 | Mixed operation | +, −, ×, ( ) |
| | Thu | 27 | 28 | Mixed operation | +, −, ÷, ( ) |
| | Fri | 29 | 30 | Mixed operation | Review |

| Week 4 | Day | Tiger Session | | Topic | Goal |
|---|---|---|---|---|---|
| | Mon | 31 | 32 | Mixed operation | +, −, ×, ÷ |
| | Tue | 33 | 34 | Mixed operation | +, −, ×, ÷, ( ) |
| | Wed | 35 | 36 | Mixed operation | +, −, ×, ÷, ( ) |
| | Thu | 37 | 38 | Mixed operation | +, −, ×, ÷, ( ), { } |
| | Fri | 39 | 40 | Mixed operation | +, −, ×, ÷, ( ), { } |

# Week 1

This week's goal is to practice dividing a 3 digit number by a 2 digit number when the quotient is a 1 digit number.

## Tiger Session

| Monday | 1 | 2 |
|--------|---|---|
| Tuesday | 3 | 4 |
| Wednesday | 5 | 6 |
| Thursday | 7 | 8 |
| Friday | 9 | 10 |

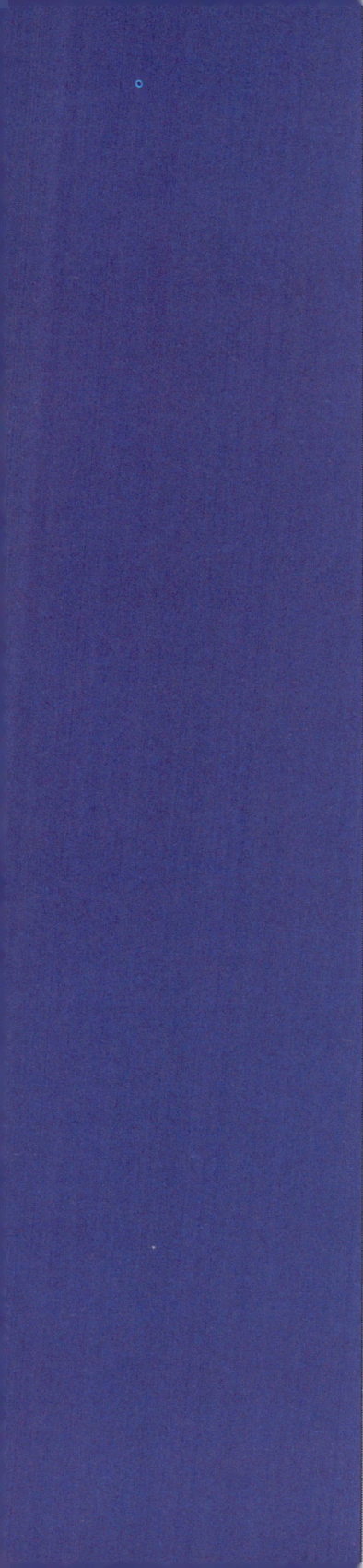

# 1

## 2 digits ÷ 2 digits ①

♠ **Divide.**

### Example

First, think about how many 21s you can put into 45.

$$21{\overline{\smash{\big)}\,4\ \ 5}}$$

*Make a rough guess with "20"as 21 is close to 20.*

Second, write 2 as a quotient as you can put 21 into 45 two times.

$$21{\overline{\smash{\big)}\,4\ \ 5}} \atop {4\ \ 2}$$

2

▲ 21 × 2

Third, subtract 42 from 45 and write 3 as a remainder.

2 R 3

$$21{\overline{\smash{\big)}\,4\ \ 5}}$$
$$\underline{4\ \ 2}$$
$$3$$

45 − 42

1)  2 R ☐

$$21{\overline{\smash{\big)}\,5\ \ 5}}$$

← 21 × 2

2)  ☐ R ☐

$$21{\overline{\smash{\big)}\,6\ \ 0}}$$

3)  ☐ R ☐

$$21{\overline{\smash{\big)}\,6\ \ 3}}$$

4)  ☐ R ☐

$$21{\overline{\smash{\big)}\,7\ \ 0}}$$

5) 22 ) 6 6    □ R □

6) 22 ) 7 0    □ R □

7) 22 ) 8 0    □ R □

8) 22 ) 8 4    □ R □

9) 22 ) 8 8    □ R □

10) 22 ) 9 3    □ R □

11) 22 ) 1 0 0    □ R □

12) 22 ) 1 0 5    □ R □

# 2

## 2 digits ÷ 2 digits ②

♠ **Divide.**

1) ☐ R ☐
   $21 \overline{)6\ 9}$

2) ☐ R ☐
   $22 \overline{)7\ 4}$

3) ☐ R ☐
   $21 \overline{)8\ 4}$

4) ☐ R ☐
   $22 \overline{)9\ 0}$

5) ☐ R ☐
   $21 \overline{)1\ 1\ 1}$

6) ☐ R ☐
   $22 \overline{)1\ 2\ 8}$

7) You bought some packs of balloons for a party. If the total number of balloons was 63, and there were 21 balloons in each pack, how many packs of balloons did you buy?

Equation: _____

Answer: _____

8) <u>Maggie and her 21 classmates</u> collected 89 falling leaves, and split them up amongst themselves. How many leaves did each person get, and how many leaves were left over?

Equation: _____

Answer:   Each person got _____ leaves, and _____ leaves were left over.

**3**  **3 digits ÷ 2 digits ①**

♠ **Divide.**

1)  $31\overline{)6\ 2}$  [ ] R [ ]

2)  $31\overline{)7\ 0}$  [ ] R [ ]

3)  $31\overline{)7\ 5}$  [ ] R [ ]

4)  $31\overline{)8\ 0}$  [ ] R [ ]

5)  $31\overline{)9\ 3}$  [ ] R [ ]

6)  $31\overline{)1\ 0\ 0}$  [ ] R [ ]

7)  $31\overline{)1\ 1\ 3}$  [ ] R [ ]

8)  $31\overline{)1\ 2\ 0}$  [ ] R [ ]

9) $32 \overline{)1\ 6\ 0}$

10) $32 \overline{)1\ 7\ 0}$

11) $32 \overline{)1\ 8\ 2}$

12) $32 \overline{)1\ 9\ 1}$

13) $32 \overline{)1\ 9\ 2}$

14) $32 \overline{)2\ 0\ 1}$

15) $32 \overline{)2\ 1\ 0}$

16) $32 \overline{)2\ 2\ 0}$

# 4

## 3 digits ÷ 2 digits ②

Date

Time spent | min

Score

♠ **Divide.**

1) $31\overline{)124}$

2) $32\overline{)96}$

3) $31\overline{)155}$

4) $32\overline{)70}$

5) $31\overline{)100}$

6) $32\overline{)224}$

7) $31\overline{)186}$

8) $32\overline{)143}$

9) I practiced the piano for 31 minutes per day. If I practiced the piano for 155 minutes in all during the last few days, for how many days did I practice the piano?

Equation: _____

Answer: _____

10) You want to divide up 98 oranges by putting 32 oranges into each box. How many boxes do you need, and how many oranges will be left over?

Equation: _____

Answer: You need _____ boxes, and _____ oranges will be left over.

**5**   **3 digits ÷ 2 digits ③**

♠ **Divide.**

Make a guess with 10 instead of 12 as 12 is close to 10!

1)  $10 \overline{)\ 7\ 5}$

2)  $12 \overline{)\ 7\ 5}$

3)  $20 \overline{)\ 1\ 0\ 3}$

4)  $23 \overline{)\ 1\ 0\ 3}$

5)  $25 \overline{)\ 1\ 0\ 3}$

6)  $27 \overline{)\ 1\ 0\ 3}$

7)  $30 \overline{)\ 1\ 2\ 5}$

8)  $34 \overline{)\ 1\ 2\ 5}$

9) $35 \overline{)2\ 1\ 0}$

10) $37 \overline{)2\ 1\ 0}$

11) $40 \overline{)2\ 4\ 5}$

12) $42 \overline{)2\ 4\ 5}$

13) $45 \overline{)3\ 2\ 0}$

14) $47 \overline{)3\ 2\ 0}$

15) $50 \overline{)2\ 7\ 5}$

16) $53 \overline{)2\ 7\ 5}$

# 6 3 digits ÷ 2 digits ④

Date

Time spent: min

Score: /

♠ **Divide.**

1) $16\overline{)88}$

2) $49\overline{)167}$

3) $28\overline{)125}$

4) $43\overline{)140}$

5) $34\overline{)182}$

6) $22\overline{)142}$

7) $43\overline{)183}$

8) $29\overline{)221}$

9) I bought some bottles of yogurt. If each bottle can hold 49 milliliters and I bought a total of 196 milliliters, how many bottles of yogurt did I buy?

Equation: _____

Answer: _____

10) There are 195 beads, and you are making necklaces, using 27 beads for each necklace. How many necklaces can you make with 195 beads, and how many beads will be left over?

Equation: _____

Answer: You can make _____ necklaces, and _____ beads will be left over.

**7** | **3 digits ÷ 2 digits ⑤**

♠ **Divide.**

1) $50 \overline{)2\ 1\ 0}$

2) $52 \overline{)2\ 1\ 0}$

3) $55 \overline{)2\ 1\ 0}$

4) $58 \overline{)2\ 1\ 0}$

5) $60 \overline{)1\ 8\ 9}$

6) $61 \overline{)1\ 8\ 9}$

7) $65 \overline{)1\ 8\ 9}$

8) $67 \overline{)1\ 8\ 9}$

9) $70 \overline{)300}$

10) $72 \overline{)300}$

11) $75 \overline{)300}$

12) $79 \overline{)300}$

13) $80 \overline{)415}$

14) $83 \overline{)415}$

15) $85 \overline{)415}$

16) $88 \overline{)415}$

# 8   3 digits ÷ 2 digits ⑥

min

♠ **Divide.**

1)  $54\overline{)2\ 3\ 0}$

2)  $67\overline{)3\ 5\ 5}$

3)  $73\overline{)2\ 5\ 6}$

4)  $82\overline{)2\ 6\ 1}$

5)  $96\overline{)2\ 6\ 8}$

6)  $56\overline{)3\ 7\ 0}$

7)  $63\overline{)4\ 8\ 5}$

8)  $78\overline{)5\ 5\ 6}$

9) Ryan jumps rope 87 times a day. If he jumped rope 261 times for a certain number of days, for how many days did he jump rope?

Equation: _____

Answer: _____

10) A teacher wants to evenly distribute 400 notebooks to 76 children. How many books will each child get, and how many books will remain?

Equation: _____

Answer: Each child will get _____ notebooks, anc _____ notebooks will remain.

**9** **3 digits ÷ 2 digits ⑦**

♠ **Divide.**

1) $35 \overline{)2\ 3\ 2}$

2) $53 \overline{)2\ 7\ 9}$

3) $72 \overline{)5\ 9\ 6}$

4) $29 \overline{)1\ 2\ 6}$

5) $48 \overline{)1\ 5\ 9}$

6) $16 \overline{)9\ 0}$

7) $84 \overline{)6\ 1\ 8}$

8) $67 \overline{)2\ 8\ 8}$

9) $55 \overline{)405}$

10) $73 \overline{)315}$

11) $20 \overline{)115}$

12) $42 \overline{)147}$

13) $86 \overline{)176}$

14) $34 \overline{)215}$

15) $97 \overline{)326}$

16) $68 \overline{)373}$

Date

Time spent | Score

min

# 10 3 digits ÷ 2 digits ⑧

♠ **Divide.**

1) $24\overline{)134}$

2) $42\overline{)267}$

3) $63\overline{)273}$

4) $16\overline{)139}$

5) $31\overline{)230}$

6) $85\overline{)283}$

7) $70\overline{)220}$

8) $58\overline{)300}$

9) At Walnut Elementary School, there are 330 students and 15 classes in 4$^{th}$ to 6$^{th}$ grade. If there are the same numbers of students in each class, how many students are there in each class?

Equation: _____

Answer: _____

10) You have 300 pieces of candy and want to put 48 pieces of candy in a pack. How many packs of candy can you make, and how many pieces of candy will remain?

Equation: _____

Answer: You can make _____ packs of candy, ard _____ pieces of candy will remain.

# Week 2

This week's goal is to practice dividing a 4 digit number by a 2 digit number when the quotient is a 2 digit number.

## Tiger Session

| Monday | 11 | 12 |
| Tuesday | 13 | 14 |
| Wednesday | 15 | 16 |
| Thursday | 17 | 18 |
| Friday | 19 | 20 |

This week, you are going to learn about how to divide 3 digits by a 2 digit divisor particularly when the quotient becomes 2 digits. It's not that difficult if you do it step by step as below.

First, write 2 as a quotient as you can put 21 into 49 TWO times.

$$
\begin{array}{r}
2\phantom{\,9\,1} \\
21)\overline{4\ 9\ 1} \\
4\ 2 \quad \leftarrow 21 \times 2 \\
\hline
\end{array}
$$

CAUTION: Align "2" and "49" to the right in the same vertical column.

Second, subtract 42 from 49 and write 7 as a remainder.

$$
\begin{array}{r}
2\phantom{\,9\,1} \\
21)\overline{4\ 9\ 1} \\
4\ 2\phantom{\,1} \\
\hline
7\phantom{\,1}
\end{array}
$$

49 − 42

Third, bring down the 1 from the dividend 491 and write next to 7.

$$
\begin{array}{r}
2\phantom{\,9\,1} \\
21)\overline{4\ 9\ 1} \\
4\ 2\phantom{\,1} \\
\hline
7\ 1
\end{array}
$$

Fourth, think about how many 21s you can put into 71.

$$
\begin{array}{r}
2\ 3 \\
21)\overline{4\ 9\ 1} \\
4\ 2\phantom{\,1} \\
\hline
7\ 1
\end{array}
$$

Fifth, write 3 as part of the quotient next to 2 as you can put 21 into 71 THREE times.

$$
\begin{array}{r}
2\ 3 \\
21)\overline{4\ 9\ 1} \\
4\ 2\phantom{\,1} \\
\hline
7\ 1 \\
21 \times 3 \rightarrow 6\ 3 \\
\hline
\end{array}
$$

Sixth, subtract 63 from 71 and write 8 as a remainder.

$$
\begin{array}{r}
2\ 3 \\
21)\overline{4\ 9\ 1} \\
4\ 2\phantom{\,1} \\
\hline
7\ 1 \\
6\ 3 \\
\hline
8
\end{array}
$$

# 11

**3 digits ÷ 2 digits ⑨**

♠ **Divide.**

1)
```
        1    R
  21) 2  4  1
       2  1
```

2)
```
           R
  21) 2  5  7
```

3)
```
           R
  21) 4  4  9
```

4)
```
           R
  21) 4  7  3
```

5)
```
           R
  21) 6  4  0
```

6)
```
           R
  21) 6  6  8
```

7) $22 \overline{)254}$ R

8) $22 \overline{)294}$ R

9) $22 \overline{)500}$ R

10) $22 \overline{)540}$ R

11) $22 \overline{)700}$ R

12) $22 \overline{)736}$ R

## 12    3 digits ÷ 2 digits ⑩

♠ **Divide.**

1) 21)3 2 0    R

2) 22)5 6 5    R

3) 21)4 5 5    R

4) 22)6 0 0    R

5) 21)6 7 7    R

6) 22)6 9 1    R

7) You have worked out 525 minutes over 21 days. How many minutes did you work out per day on average?

Equation:

Answer:

8) There are 24 students in a classroom, and the teacher wants to evenly divide up 763 pieces of candy for the students. How many pieces of candy can the teacher give to each student, and how many pieces of candy will be left over?

Equation:

Answer:

**13**  **3 digits ÷ 2 digits ⑪**

♠ **Divide.**

1)  31)3 4 9

2)  31)3 5 5

3)  31)4 1 0

4)  31)4 2 5

5)  31)6 6 2

6)  31)9 9 8

7) $32\overline{)3\ 6\ 2}$

8) $32\overline{)3\ 7\ 0}$

9) $32\overline{)4\ 5\ 5}$

10) $32\overline{)4\ 6\ 2}$

11) $32\overline{)7\ 1\ 0}$

12) $32\overline{)7\ 5\ 2}$

Date _____

Time spent [ min ]   Score [ / ]

♠ **Divide.**

1)  31 ) 3 8 3

2)  32 ) 4 6 3

3)  31 ) 7 2 3

4)  32 ) 8 0 7

5)  31 ) 9 6 7

6)  32 ) 9 6 1

7) A printer can print out 31 pages per minute. How many minutes will it take to print out 744 pages?

Equation: _____

Answer: _____

8) In a store, there are 534 cans and 32 empty boxes. You want to equally divide up the cans to put them into 32 boxes. How many cans will you be able to put into each box, and how many cans will remain?

Equation: _____

Answer: _____

♠ **Divide.**

1) 20)1 1 2 1    R

2) 21)1 1 2 1    R

3) 23)1 1 2 1    R

4) 30)1 3 0 0    R

5) 28)1 3 0 0    R

6) 32)1 3 0 0    R

7) $34 \overline{)1\ 3\ 0\ 0}$ R

8) $38 \overline{)1\ 3\ 0\ 0}$ R

9) $40 \overline{)2\ 0\ 5\ 0}$ R

10) $42 \overline{)2\ 0\ 5\ 0}$ R

11) $43 \overline{)2\ 0\ 5\ 0}$ R

12) $48 \overline{)2\ 0\ 5\ 0}$ R

## 16   4 digits ÷ 2 digits ②

♠ **Divide.**

1)  $23 \overline{) 1\,1\,9\,0}$

2)  $38 \overline{) 1\,7\,9\,0}$

3)  $41 \overline{) 1\,4\,9\,0}$

4)  $34 \overline{) 1\,2\,4\,3}$

5)  $27 \overline{) 1\,2\,2\,5}$

6)  $16 \overline{) 1\,1\,6\,0}$

7) How many hours are there in 2280 minutes?

Equation: _____

Answer: _____

8) You need 15 oranges to make 1 liter of orange juice. How many liters of orange juice can you make with 1260 oranges?

Equation: _____

Answer: _____

**17**   **4 digits ÷ 2 digits ③**

♠ **Divide.**

1)  50) 1 3 1 0

2)  52) 1 3 1 0

3)  57) 1 3 1 0

4)  60) 2 2 6 0

5)  63) 2 2 6 0

6)  68) 2 2 6 0

7) $70 \overline{)2\ 5\ 0\ 0}$

8) $72 \overline{)2\ 5\ 0\ 0}$

9) $80 \overline{)2\ 2\ 0\ 0}$

10) $78 \overline{)2\ 2\ 0\ 0}$

11) $90 \overline{)3\ 7\ 9\ 0}$

12) $93 \overline{)3\ 7\ 9\ 0}$

# 18 4 digits ÷ 2 digits ④

Date

| Time spent | Score |
| --- | --- |
| min | / |

♠ **Divide.**

1) $50\overline{)2170}$

2) $53\overline{)2170}$

3) $60\overline{)2770}$

4) $57\overline{)2770}$

5) $59\overline{)2770}$

6) $63\overline{)2770}$

7) Jackson earned 1335 dollars by selling 89 pepperoni pizzas. How much does a pepperoni pizza cost?

Equation: _____

Answer: _____

8) There are 2870 inches of packaging rope. You want to cut the rope to equal length for packaging up the boxes. If each cut of packaging rope is 51 inches, how many pieces of packaging rope can you cut, and how many inches will be left over?

Equation: _____

Answer: _____

# 19   4 digits ÷ 2 digits ⑤

♠ **Divide.**

1)  $51 \overline{)1\ 1\ 7\ 7}$

2)  $29 \overline{)1\ 1\ 4\ 2}$

3)  $76 \overline{)1\ 9\ 2\ 2}$

4)  $42 \overline{)6\ 7\ 7}$

5)  $35 \overline{)1\ 7\ 0\ 0}$

6)  $68 \overline{)1\ 6\ 1\ 0}$

7) $49\overline{)795}$

8) $64\overline{)2258}$

9) $83\overline{)1759}$

10) $32\overline{)588}$

11) $50\overline{)3825}$

12) $77\overline{)3033}$

**20**   **4 digits ÷ 2 digits ⑥**

♠ **Divide.**

1) $51 \overline{)1\ 7\ 7\ 7}$

2) $37 \overline{)1\ 4\ 6\ 0}$

3) $23 \overline{)1\ 1\ 1\ 1}$

4) $49 \overline{)1\ 0\ 0\ 0}$

5) $62 \overline{)1\ 1\ 8\ 8}$

6) $18 \overline{)1\ 0\ 7\ 0}$

7) Chris works out for 43 minutes a day. If he worked out for a total of 1333 minutes, how many days did he work out?

Equation:

Answer:

8) How many dozen eggs can you make with 1876 eggs, and how many eggs will be left over?

Equation:

Answer:

# Week 3

This week's goal is to practice mixed operations with addition, subtraction, multiplication, division, and brackets in the range of natural numbers.

## Tiger Session

| Monday | 21 | 22 |
| Tuesday | 23 | 24 |
| Wednesday | 25 | 26 |
| Thursday | 27 | 28 |
| Friday | 29 | 30 |

**21** **Mixed Operations** ①
$+, -, ( \ )$

♠ **Solve the problems.**

1) $25 + 10 - 13 =$

For + and −:
**FROM LEFT TO RIGHT!**

2) $48 + 17 - 12 =$

3) $29 - 14 + 18 =$

4) $54 - 26 - 12 =$

5) $35 + 46 - 21 =$

6) $42 - 12 + 25 =$

7) $24 + (11 - 7) =$

BRACKETS ALWAYS FIRST!

8) $36 - 21 - 10 =$

9) $42 - 12 + 20 =$

10) $54 - 20 - 7 =$

See the differing results!

11) $54 - (20 - 7) =$

12) $45 + 10 - 25 =$

13) $37 + (21 - 15) =$

**22** **Mixed Operations** ②
+, −, ( )

♠ **Solve the problems.**

1) $27 + 23 - 15 =$

2) $36 - 24 + 12 - 3 =$

3) $22 - (15 + 5) =$

4) $22 - 15 + 5 =$

5) $40 + 19 - 24 =$

6) $56 - (23 - 11) + 4 =$

7) Joe runs a shoe store. Yesterday, there were 36 pairs of shoes in the store, and 24 pairs were sold. Today, he brought in 20 pairs of shoes to stock up. How many pairs of shoes are there in the stock now?

Equation: _____

Answer: _____

8) There were 24 birds singing in a tree. After a while, 30 more birds came to the tree, and 41 birds flew away. Then how many birds are left in the tree?

Equation: _____

Answer: _____

**23** **Mixed Operations** ③

$\times, \div, (\ )$

♠ **Solve the problems.**

1) $5 \times 8 \div 2 =$

For $\times$ and $\div$:
**FROM LEFT TO RIGHT!**

2) $10 \div 5 \times 7 =$

3) $12 \times 3 \div 4 =$

4) $3 \times 5 \times 6 =$

5) $7 \times 10 \div 2 =$

6) $20 \div 2 \div 5 =$

7)  $6 \times (8 \div 2) =$

8)  $48 \div 8 \times 9 =$

9)  $72 \div (3 \times 3) =$

10)  $10 \times 11 \div 5 =$

11)  $(80 \div 4) \times 3 =$

12)  $49 \div 7 \times 5 =$

13)  $64 \div (32 \div 4) =$

**24**    **Mixed Operations** ④
        $\times, \div, (\ )$

min

♠ **Solve the problems.**

1)  $5 \times 3 \times 7 =$

2)  $54 \div 9 \times 11 =$

3)  $40 \times 2 \div 8 \div 2 =$

4)  $88 \div 4 \times 10 =$

5)  $55 \div 5 \times 9 =$

6)  $72 \div 9 \times 20 \div 4 =$

7) There are three pizzas on a table, and mom cut each pizza into 6 pieces. Then she equally distributed all the pieces for 9 kids. How many pieces did each kid receive?

Equation: _____

Answer: _____

8) You bought 6 packs of balloons, each of which has 8 balloons. Then you equally divided the total balloons for your 7 friends. How many balloons did you give to each of your 7 friends? How many balloons were left?

Equation: _____

Answer: _____

# 25 Mixed Operations ⑤

$+, -, \times, (\ )$

♠ **Solve the problems.**

> REMEMBER the calculation order:
>
> Brackets → ×, ÷ → +, −
>
> From LEFT to RIGHT    From LEFT to RIGHT

1) $4 + 2 \times 5 - 3 =$

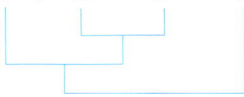

See how the results differ!

2) $(4 + 2) \times 5 - 3 =$

3) $6 - 2 + 3 \times 5 =$

4) $36 - (2 + 3) \times 5 =$

5) $5 \times 3 - 2 \times 4 =$

6) $18 + 4 - 3 \times 7 =$

7) $18 + (4 - 3) \times 7 =$

8) $27 - 8 + 4 \times 2 =$

9) $27 - (8 + 4) \times 2 =$

10) $5 \times 2 + 3 - 9 =$

11) $5 \times (2 + 3) - 9 =$

12) $15 + 5 - (4 + 6) =$

**26**

**Mixed Operations** ⑥

$+, -, \times, (\ )$

Date

Time spent

min

Score

♠ **Solve the problems.**

1) $9 + 2 \times 3 - 4 =$

2) $(9 + 2) \times 3 - 4 =$

3) $35 - 5 + 4 \times 3 =$

4) $35 - (5 + 4) \times 3 =$

5) $4 \times 5 + 5 \times 7 =$

6) $4 \times (5 + 5) - 7 =$

7) I had 3 stickers, and my brother had 5 times more stickers than I did. My brother and I decided to gather all the stickers and to give 11 stickers to my friend Jeffrey. After giving away the stickers, how many stickers are left?

Equation: _____

Answer: _____

8) On Monday, I borrowed 3 books. Then I borrowed 4 books on each day from Tuesday to Friday in a row. If I returned 11 books on Friday, then how many books were still left to return?

Equation: _____

Answer: _____

**27** **Mixed Operations** ⑦
+, −, ÷, ( )

min

♠ **Solve the problems.**

REMEMBER the calculation order:

Brackets ⟶ ×, ÷ ⟶ +, −

From LEFT to RIGHT | From LEFT to RIGHT

1) $18 - 6 \div 2 + 3 =$

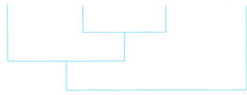

See how the results differ!

2) $(18 - 6) \div 2 + 3 =$

3) $42 \div 3 + 4 \div 2 =$

4) $42 \div (3 + 4) \div 2 =$

5) $28 - 8 \div 4 + 10 =$

6) $(28 - 8) \div 4 + 10 =$

7) $10 + 25 - 20 \div 5 =$

8) $10 + (25 - 20) \div 5 =$

9) $36 \div 9 - 3 + 8 =$

10) $36 \div (9 - 3) + 8 =$

11) $30 - 24 \div 6 + 2 =$

12) $30 - 24 \div (6 + 2) =$

## 28 Mixed Operations ⑧

$+, -, \div, (\ )$

♠ **Solve the problems.**

1) $15 + 5 \div 5 - 4 =$

2) $(15 + 5) \div 5 - 4 =$

3) $48 \div 8 - 4 + 7 =$

4) $48 \div (8 - 4) + 7 =$

5) $20 - 4 + 16 \div 2 =$

6) $20 - (4 + 16) \div 2 =$

7) There are 48 Jellybeans on a table. After eating 6 jellybeans, you divide up the jellybeans for 6 plates. Then you add 5 jellybeans to each plate. How many jellybeans are on each plate?

Equation: _____

Answer: _____

8) There were 21 M&Ms in a box, and you added 11 more M&Ms to the box. After taking away 17 M&Ms, you equally divided up M&Ms into 5 Ziploc bags. How many M&Ms are in each bag?

Equation: _____

Answer: _____

**29** **Mixed Operations** ⑨

Review

♠ **Solve the problems.**

1) $(10 + 5) \div 5 - 2 =$

2) $24 + 17 - (11 - 6) =$

3) $4 \times 8 - 2 \times 3 =$

4) $16 - 9 \div 3 - 5 =$

5) $(36 - 16) \div 2 + 10 =$

6) $20 \times 2 - 4 + 5 =$

7) $48 - 20 \div 5 + 4 =$

8) $18 - (17 - 10) + 15 =$

9) $7 \times (8 + 2) - 18 =$

10) $33 - 10 \times 3 + 7 =$

11) $25 - 5 + 4 \times 2 =$

12) $(16 - 4) \div 6 + 15 =$

13) $42 + 8 \times 4 - 13 =$

**30** **Mixed Operations** ⑩

Review

♠ **Solve the problems.**

1) $7 \times 7 - 3 \times 3 =$

2) $32 - 24 \div 8 - 10 =$

3) $20 \times (8 - 6) + 8 =$

4) $17 - 8 + 28 \div 7 =$

5) $(20 + 9) - 7 \times 4 =$

6) $40 - 30 \div 3 + 12 =$

7) From the library, I borrowed 3 books, and my brother borrowed 7 books per week for the past 5 weeks in a row. How many more books did my brother borrow than I did during the last 5 weeks?

Equation: _____

Answer: _____



8) You made 8 origami birds, and your sister made 5 origami birds every day for the last 3 days. Out of the total, you gave 21 origami birds to your mom. How many origami birds are left?

Equation: _____

Answer: _____

# Week 4

This week's goal is to practice mixed operations with addition, subtraction, multiplication, division, and brackets in the range of natural numbers.

## Tiger Session

| | | |
|---|---|---|
| **Monday** | 31 | 32 |
| **Tuesday** | 33 | 34 |
| **Wednesday** | 35 | 36 |
| **Thursday** | 37 | 38 |
| **Friday** | 39 | 40 |

♠ **Solve the problems.**

REMEMBER the calculation order:

Brackets ➜ ×, ÷ ➜ +, −

From LEFT to RIGHT    From LEFT to RIGHT

1) $3 \times 2 - 8 \div 4 =$

2) $15 - 12 \div 4 \times 5 =$

3) $15 - 12 \div 4 \times 5 =$

4) $42 \div 7 + 5 \times 4 =$

5) $10 + 28 \div 4 \times 10 =$

6) $9 \times 6 \div 3 - 12 =$

7) $12 + 8 \times 2 - 24 =$

8) $10 + (25 - 20) \div 5 =$

9) $45 \div 9 + 6 \times 8 =$

10) $24 \div 3 \times 5 - 20 =$

11) $33 - 20 \times 3 \div 6 =$

12) $12 \times 6 - 63 \div 9 =$

**32** **Mixed Operations** ⑫
+, −, ×, ÷

♠ **Solve the problems.**

1) $45 \div 5 \times 6 + 13 =$

2) $21 - 8 \div 4 \times 7 =$

3) $7 \times 6 + 56 \div 8 =$

4) $81 \div 9 + 12 \times 3 =$

5) $6 \times 8 - 48 \div 6 =$

6) $36 \div 3 - 6 \times 2 =$

7) Yesterday you practiced piano for 21 minutes, and today 18 more minutes than you did yesterday. Your sister also practiced piano today 19 less minutes than you did today. And your brother practiced piano today 3 times more than your sister. How many minutes did you brother practice piano today?

Equation: _____

Answer: _____

8) You had 98 dollars in your savings account. You added 8 dollars to the savings account each day for the last five days. Then you withdrew 24 dollars from the account. How many dollars do you have in the account now?

Equation: _____

Answer: _____

**33** **Mixed Operations** ⑬
$+, -, \times, \div, (\ )$

♠ Solve the problems.

1) $21 + 3 \times 4 - 24 \div 6 =$

2) $(40 - 12) \div 2 + 5 \times 3 =$

3) $11 + 18 \div 3 \times 5 - 10 =$

4) $6 \times 7 \div (5 - 3) + 15 =$

5) $72 \div 9 + 3 \times 11 - 20 =$

6) $54 - 24 \div 6 \times (8 - 5) =$

7) $4 \times 9 - (42 \div 7) + 16 =$

8) $48 \div 8 \times (9 - 7) \times 6 =$

9) $27 - 5 \times 3 - 22 \div 2 =$

10) $36 \div (8 - 4) \times 5 - 7 =$

11) $6 \times 9 \div (15 - 9) - 9 =$

12) $46 - 8 \times 5 + 24 \div 2 =$

# 34 Mixed Operations ⑭

$+, -, \times, \div, (\ )$

min

♠ **Solve the problems.**

1)  $21 + 3 \times 4 - 24 \div 6 =$

2)  $(40 - 12) \div 2 + 5 \times 3 =$

3)  $11 + 18 \div 3 \times 5 - 10 =$

4)  $6 \times 7 \div (5 - 3) + 15 =$

5)  $72 \div 9 + 3 \times 11 - 20 =$

6)  $54 - 24 \div 6 \times (8 - 5) =$

7) I am 8 years old, and my mom is 4 times older than me. My brother is 7 years old, and my grandma is 9 times older than my brother. How much older is my grandma than mom?

Equation: _____

Answer: _____

8) You had 96 Pokémon cards, and yesterday mom bought 28 more cards for me. Then you equally distributed the cards to 4 of your brothers. How many cards did each of your brothers get?

Equation: _____

Answer: _____

**35** **Mixed Operations** ⑮
$+, -, \times, \div, (\ )$

min

♠ **Solve the problems.**

1) $35 - 28 \div 4 \times 2 + 10 =$

2) $(26 + 12) \div 2 + 6 \times 9 =$

3) $54 \div 3 + (17 - 15) \times 9 =$

4) $6 \times 4 - 88 \div (4 \times 2) =$

5) $25 \times (8 \div 2) + 19 - 8 =$

6) $69 - 8 \times 11 \div (21 - 17) =$

7) $36 \div (16 - 10) + 3 \times 5 =$

8) $24 + 32 \div 8 \times 7 - 9 =$

9) $21 \div (3 + 4) \times (25 - 15) =$

10) $(15 - 10) + 7 \times 6 - 8 \div 2 =$

11) $63 \div 3 \times (28 - 24) + 13 =$

12) $37 - 25 \div 5 \times (15 - 11) + 6 =$

**36**  **Mixed Operations** ⑯
$+, -, \times, \div, ( \ )$

min

♠ **Solve the problems.**

1)  $6 \times (5 + 5) - 31 + 42 \div 2 =$

2)  $(29 - 9) \div 5 \times (7 + 3) - 6 =$

3)  $36 - (6 + 5) \times 3 - 21 \div 7 =$

4)  $72 \div (12 - 3) \times 2 - 14 =$

5)  $(15 + 5) \div (26 - 21) \times 8 =$

6)  $8 \times 9 \div (12 - 8) + 27 =$

7) There are a certain number of boys and 8 girls in my class. The teacher wants to give 3 pencils to each student, so she prepares 60 pencils. If no pencils are left after distributing, how many boys are in the class?

Equation: _____

Answer: _____

8) Red roses were planted in 5 rows, and in each row, there were 4 red roses. On the other side, yellow roses were planted in 7 rows, and there were 3 yellow roses in each row. One day, 9 red roses and 11 yellow roses froze due to the frost. How many more red roses were left than yellow roses?

Equation: _____

Answer: _____

**37** **Mixed Operations** ⑰
$+, -, \times, \div, ( \ ), \{ \ \}$

♠ Solve the problems.

1) $\{(3 \times 5) - 8\} + 9 =$

2) $27 - \{15 + (24 \div 8)\} =$

3) $\{38 - (4 \times 7)\} - 6 =$

4) $\{(12 \div 3) - 3\} + 8 =$

5) $\{(2 \times 9) + 2\} \div 5 =$

6) $\{(2 \times 5) - 5\} \times 3 =$

7) $10 - \{(3 \times 4) - (9 \div 3)\} =$

8) $\{(6 \times 3) + 5\} - (72 \div 8) =$

9) $36 \div \{32 - (4 \times 7)\} =$

10) $\{(27 \div 3) + 4 - 8\} \times 5 =$

11) $25 - \{28 - (2 \times 7) + 10\} =$

12) $\{(36 - 6) \div 5 + 4\} - 10 =$

**38** **Mixed Operations** ⑱
$+, -, \times, \div, (\ \ ), \{\ \ \}$

♠ **Solve the problems.**

1) $\{22 - (7 \times 3)\} + 9 =$

2) $26 + \{25 \div 5 \times (6 - 3)\} =$

3) $2 \times \{(39 \div 3) - (26 - 20)\} =$

4) $\{10 \div (8 - 6) \times 5\} - 8 =$

5) $53 - \{(4 \times 9) - (64 \div 8)\} =$

6) $\{34 - (4 + 14) \div 2\} - 12 =$

7) There were 7 boxes in each of which there are 10 cookies. You unpacked all the boxes and gathered all the cookies. After giving 32 cookies to your brother, you equally divided the cookies left to put them into 5 zip-loc bags. How many cookies did you put into each zip-loc bag, and how many cookies were left?

Equation:  _____

Answer:  _____

8) You went apple-picking with your mom and dad. You picked 8 apples, dad picked apples 5 more times than you, and mom picked 9 more apples than you. How many apples dic your family pick in total?

Equation:  _____

Answer:  _____

**39**  **Mixed Operations** ⑲
$+, -, \times, \div, (\ ), \{\ \}$

min

♠ **Solve the problems.**

1)  $36 \div 4 + \{(3 + 9) - 10\} + 8 =$

2)  $\{27 \div 3 \times (15 - 13)\} + 2 \times 3 =$

3)  $(41 - 39) \times 20 - \{(25 + 5) \div 3 \times 4\} =$

4)  $\{55 \div (11 - 6) \times 3\} - (13 - 8) =$

5)  $\{(16 - 5) \times (4 + 3)\} + 18 \div 3 =$

6) $33 + \{(12 - 6) \times 3 \div 9\} - 29 =$

7) $\{(28 \div 2) \times 3 \div 7\} + (54 - 37) =$

8) $(13 + 27) \div \{38 - (4 \times 9)\} =$

9) $78 - 35 - \{22 \div (16 - 14) \times 3\} =$

10) $\{(42 \div 3) \div (27 - 20)\} + 25 =$

**40** **Mixed Operations** ⑳

$+, -, \times, \div, (\ \ ), \{\ \ \}$

♠ **Solve the problems.**

1)  $\{(26 + 16) \div 3 - 8\} \times 3 =$

2)  $50 - \{36 - (56 + 44) \div 5\} =$

3)  $\{(20 - 10) \times (56 \div 7)\} - (16 - 6) =$

4)  $86 + \{(20 \times 2) \div 8 + 18\} - 8 =$

5)  $(21 - 15) \times \{36 - (45 \div 5) \times 3\} =$

6) Last week, I ran 27 minutes every day from Monday to Friday.
This week, I ran 37 minutes every day from Monday to Thursday.
In which one between last week and this week, did I run more?
And how much more?

Equation:

Answer:

7) There are 99 marbles in a big jar. After taking out 7 marbles 8
times from the jar in a row, you added 18 marbles to the jar.
Then you equally divide the total marbles into 6 plates. How
many marbles did you put into each plate, and how many
marbles were left?

Equation:

Answer:

# F – 1: Answers

# Week 1

## 1 (p. 5 ~ 6)

① 21)5 5 → 2 R 13; 4 2; 1 3
② 21)6 0 → 2 R 18; 4 2; 1 8
③ 21)6 3 → 3 R 0; 6 3; 0
④ 21)7 0 → 3 R 7; 6 3; 7
⑤ 22)6 6 → 3 R 0; 6 6; 0 0
⑥ 22)7 0 → 3 R 4; 6 6; 4
⑦ 22)8 0 → 3 R 14; 6 6; 1 4
⑧ 22)8 4 → 3 R 18; 6 6; 1 8
⑨ 22)8 8 → 4 R 0; 8 8; 0
⑩ 22)9 3 → 4 R 5; 8 8; 5
⑪ 22)1 0 0 → 4 R 12; 8 8; 1 2
⑫ 22)1 0 5 → 4 R 17; 8 8; 1 7

## 2 (p. 7 ~ 8)

① 21)6 9 → 3 R 6; 6 3; 6
② 22)7 4 → 3 R 8; 6 6; 8
③ 21)8 4 → 4 R 0; 8 4; 0
④ 22)9 0 → 4 R 2; 8 8; 2
⑤ 21)1 1 1 → 5 R 6; 1 0 5; 6
⑥ 22)1 2 8 → 5 R 18; 1 1 0; 1 8

⑦ 63 ÷ 21 = 3, 3 packs
⑧ 89 ÷ 21 = 4 R 1, 4, 1

## 3 (p. 9 ~ 10)

| ① 2R0 | ② 2R8 | ③ 2R13 | ④ 2R18 |
| ⑤ 3R0 | ⑥ 3R7 | ⑦ 3R20 | ⑧ 3R27 |
| ⑨ 5R0 | ⑩ 5R10 | ⑪ 5R22 | ⑫ 5R31 |
| ⑬ 6R0 | ⑭ 6R9 | ⑮ 6R18 | ⑯ 6R28 |

## 4 (p. 11 ~ 12)

| ① 4R0 | ② 3R0 | ③ 5R0 | ④ 2R6 |
| ⑤ 3R7 | ⑥ 7R0 | ⑦ 6R0 | ⑧ 4R15 |

⑨ 155 ÷ 31 = 5, 5 days   ⑩ 98 ÷ 32 = 3R2, 3, 2

## 5 (p. 13 ~ 14)

| ① 7R5 | ② 6R3 | ③ 5R3 | ④ 4R11 |
| ⑤ 4R3 | ⑥ 3R22 | ⑦ 4R5 | ⑧ 3R23 |
| ⑨ 6R0 | ⑩ 5R25 | ⑪ 6R5 | ⑫ 5R35 |
| ⑬ 7R5 | ⑭ 6R38 | ⑮ 5R25 | ⑯ 5R10 |

## 6 (p. 15 ~ 16)

| ① 5R8 | ② 3R20 | ③ 4R13 | ④ 3R11 |
| ⑤ 5R12 | ⑥ 6R10 | ⑦ 4R11 | ⑧ 7R18 |

⑨ 196 ÷ 49 = 4, 4 bottles
⑩ 195 ÷ 27 = 7R6, 7, 6

## 7 (p. 17 ~ 18)

| ① 4R10 | ② 4R2 | ③ 3R45 | ④ 3R36 |
| ⑤ 3R9 | ⑥ 3R6 | ⑦ 2R59 | ⑧ 2R55 |
| ⑨ 4R20 | ⑩ 4R12 | ⑪ 4R0 | ⑫ 3R63 |
| ⑬ 5R15 | ⑭ 5R0 | ⑮ 4R75 | ⑯ 4R63 |

## 8 (p. 19 ~ 20)

| ① 4R14 | ② 5R20 | ③ 3R37 | ④ 3R15 |
| ⑤ 2R76 | ⑥ 6R34 | ⑦ 7R44 | ⑧ 7R10 |

⑨ 261 ÷ 87 = 3, 3 days
⑩ 400 ÷ 76 = 5R20, 5, 20

## 9 (p. 21 ~ 22)

| ① 6R22 | ② 5R14 | ③ 8R20 | ④ 4R10 |
| ⑤ 3R15 | ⑥ 5R10 | ⑦ 7R30 | ⑧ 4R20 |
| ⑨ 7R20 | ⑩ 4R23 | ⑪ 5R15 | ⑫ 3R21 |
| ⑬ 2R4 | ⑭ 6R11 | ⑮ 3R35 | ⑯ 5R33 |

| 10 | (p. 23 ~ 24) |
|----|----|

① 5R14  ② 6R15  ③ 4R21  ④ 8R11
⑤ 7R13  ⑥ 3R28  ⑦ 3R10  ⑧ 5R10
⑨ 330 ÷ 15 = 22,  22 students
⑩ 300 ÷ 48 = 6R12,  6, 12

## Week 2

| 11 | (p. 27 ~ 28) |
|----|----|

① 11R10  ② 12R5  ③ 21R8  ④ 22R11
⑤ 30R10  ⑥ 31R17  ⑦ 11R12  ⑧ 13R8
⑨ 22R16  ⑩ 24R12  ⑪ 31R18  ⑫ 33R10

| 12 | (p. 29 ~ 30) |
|----|----|

① 15R5  ② 25R15  ③ 21R14  ④ 27R6
⑤ 32R5  ⑥ 31R9
⑦ 525 ÷ 21 = 25,  25 minutes
⑧ 763 ÷ 24 = 31R19,  The teacher can give 31 pieces of candy to each student, and 19 pieces will be left over.

| 13 | (p. 31 ~ 32) |
|----|----|

① 11R8  ② 11R14  ③ 13R7  ④ 13R22
⑤ 21R11  ⑥ 32R6  ⑦ 11R10  ⑧ 11R18
⑨ 14R7  ⑩ 14R14  ⑪ 22R6  ⑫ 23R16

| 14 | (p. 33 ~ 34) |
|----|----|

① 12R11  ② 14R15  ③ 23R10  ④ 25R7
⑤ 31R6  ⑥ 30R1
⑦ 744 ÷ 31 = 24,  24 minutes
⑧ 532 ÷ 32 = 16R20, You will be able to put 16 cans into each box, and 20 cans will remain.

| 15 | (p. 35 ~ 36) |
|----|----|

① 56R1  ② 53R8  ③ 48R17  ④ 43R10
⑤ 46R12  ⑥ 40R20  ⑦ 38R8  ⑧ 34R8
⑨ 51R10  ⑩ 48R34  ⑪ 47R29  ⑫ 42R34

| 16 | (p. 37 ~ 38) |
|----|----|

① 51R17  ② 47R4  ③ 36R14  ④ 36R19
⑤ 45R10  ⑥ 72R8
⑦ 2280 ÷ 60 = 38,  38 hours
⑧ 1260 ÷ 15 = 84,  84 liters

| 17 | (p. 39 ~ 40) |
|----|----|

① 26R10  ② 25R10  ③ 22R56  ④ 37R40
⑤ 35R55  ⑥ 33R16  ⑦ 35R50  ⑧ 34R52
⑨ 27R40  ⑩ 28R16  ⑪ 42R10  ⑫ 40R70

| 18 | (p. 41 ~ 42) |
|----|----|

① 43R20  ② 40R50  ③ 46R10  ④ 48R34
⑤ 46R56  ⑥ 43R61
⑦ 1335 ÷ 89 = 15,  15 dollars
⑧ 2870 ÷ 51 = 56R14
You can cut 56 pieces of packaging rope, and 14 inches will be left over.

| 19 | (p. 43 ~ 44) |
|----|----|

① 23R4  ② 39R11  ③ 25R22  ④ 16R5
⑤ 48R20  ⑥ 23R46  ⑦ 16R11  ⑧ 35R18
⑨ 21R16  ⑩ 18R12  ⑪ 76R25  ⑫ 39R30

| 20 | (p. 45 ~ 46) |
|----|----|

① 34R43  ② 39R17  ③ 48R7  ④ 20R20
⑤ 19R10  ⑥ 59R8
⑦ 1333 ÷ 43 = 31,  31 days
⑧ 1876 ÷ 12 = 156R4, You can make 156 dozen of eggs, and 4 eggs will be left over.

## Week 3

| 21 | (p. 49 ~ 50) |
|----|----|

① 22  ② 53  ③ 33  ④ 16  ⑤ 60
⑥ 55  ⑦ 28  ⑧ 5  ⑨ 50  ⑩ 27
⑪ 41  ⑫ 30  ⑬ 43

| 22 | (p. 51 ~ 52) |
|----|----|

① 35  ② 21  ③ 2  ④ 12  ⑤ 35
⑥ 48  ⑦ 36 − 24 + 20 = 32, 32 pairs
⑧ 24 + 30 − 41 = 13, 13 birds

| 23 | (p. 53 ~ 54) |
|----|----|

① 20  ② 14  ③ 9  ④ 90  ⑤ 35
⑥ 2  ⑦ 24  ⑧ 54  ⑨ 8  ⑩ 22
⑪ 60  ⑫ 35  ⑬ 8

| 24 | (p. 55 ~ 56) |
|----|----|

① 105  ② 66  ③ 5  ④ 220  ⑤ 99
⑥ 40  ⑦ 3 x 6 ÷ 9 = 2, 2 pieces

$6 \times 8 \div 7 = 6R6$

⑧ You gave 6 balloons to your each of your friends, and 6 balloons were left.

| **25** | (p. 57 ~ 58) | | | |
|---|---|---|---|---|
| ① 11 | ② 27 | ③ 19 | ④ 11 | ⑤ 7 |
| ⑥ 1 | ⑦ 25 | ⑧ 27 | ⑨ 3 | ⑩ 4 |
| ⑪ 16 | ⑫ 10 | | | |

| **26** | (p. 59 ~ 60) | | | |
|---|---|---|---|---|
| ① 11 | ② 29 | ③ 42 | ④ 8 | ⑤ 55 |
| ⑥ 33 | ⑦ $3 + 3 \times 5 - 11 = 7$, 7 stickers | | | |
| ⑧ $3 + 4 \times 4 - 11 = 8$, 8 books left | | | | |

| **27** | (p. 61 ~ 62) | | | |
|---|---|---|---|---|
| ① 18 | ② 9 | ③ 16 | ④ 3 | ⑤ 36 |
| ⑥ 15 | ⑦ 31 | ⑧ 11 | ⑨ 9 | ⑩ 14 |
| ⑪ 28 | ⑫ 27 | | | |

| **28** | (p. 63 ~ 64) | | | |
|---|---|---|---|---|
| ① 12 | ② 0 | ③ 9 | ④ 19 | ⑤ 24 |
| ⑥ 10 | ⑦ $(48 - 6) \div 6 + 5 = 12$, 12 jellybeans | | | |
| ⑧ $(21 + 11 - 17) \div 5 = 3$, 3 M&Ms | | | | |

| **29** | (p. 65 ~ 66) | | | |
|---|---|---|---|---|
| ① 1 | ② 36 | ③ 26 | ④ 8 | ⑤ 20 |
| ⑥ 41 | ⑦ 48 | ⑧ 26 | ⑨ 52 | ⑩ 10 |
| ⑪ 28 | ⑫ 17 | ⑬ 61 | | |

| **30** | (p. 67 ~ 68) | | | |
|---|---|---|---|---|
| ① 40 | ② 19 | ③ 48 | ④ 13 | ⑤ 1 |
| ⑥ 42 | ⑦ $(7 \times 5) - (3 \times 5) = 20$, 20 books | | | |
| ⑧ $(8 + 5) \times 3 - 21 = 18$, 18 origami birds | | | | |

**Week 4**

| **31** | (p. 71 ~ 72) | | | |
|---|---|---|---|---|
| ① 4 | ② 0 | ③ 16 | ④ 26 | ⑤ 80 |
| ⑥ 6 | ⑦ 4 | ⑧ 11 | ⑨ 53 | ⑩ 20 |
| ⑪ 23 | ⑫ 65 | | | |

| **32** | (p. 73 ~ 74) | | | |
|---|---|---|---|---|
| ① 67 | ② 7 | ③ 49 | ④ 45 | ⑤ 40 |
| ⑥ 0 | ⑦ $(21 + 18 - 19) \times 3 = 60$, 60 minutes | | | |
| ⑧ $98 + 8 \times 5 - 24 = 114$, 114 dollars | | | | |

| **33** | (p. 75 ~ 76) | | | |
|---|---|---|---|---|
| ① 29 | ② 29 | ③ 31 | ④ 36 | ⑤ 21 |
| ⑥ 42 | ⑦ 46 | ⑧ 72 | ⑨ 1 | ⑩ 38 |
| ⑪ 0 | ⑫ 18 | | | |

| **34** | (p. 77 ~ 78) | | | |
|---|---|---|---|---|
| ① 29 | ② 29 | ③ 31 | ④ 36 | ⑤ 21 |
| ⑥ 42 | ⑦ $7 \times 9 - 8 \times 4 = 31$, 31 years older | | | |
| ⑧ $(96 + 28) \div 4 = 31$, 31 cards | | | | |

| **35** | (p. 79 ~ 80) | | | |
|---|---|---|---|---|
| ① 31 | ② 73 | ③ 36 | ④ 13 | ⑤ 111 |
| ⑥ 47 | ⑦ 21 | ⑧ 43 | ⑨ 30 | ⑩ 43 |
| ⑪ 97 | ⑫ 23 | | | |

| **36** | (p. 81 ~ 82) | | | |
|---|---|---|---|---|
| ① 50 | ② 34 | ③ 0 | ④ 2 | ⑤ 32 |
| ⑥ 45 | ⑦ $\{60 - (8 \times 3)\} \div 3 = 12$, 12 boys | | | |
| ⑧ $(5 \times 4 - 9) - (7 \times 3 - 11) = 1$, 1 red rose | | | | |

| **37** | (p. 83 ~ 84) | | | |
|---|---|---|---|---|
| ① 16 | ② 9 | ③ 4 | ④ 9 | ⑤ 4 |
| ⑥ 15 | ⑦ 1 | ⑧ 14 | ⑨ 9 | ⑩ 25 |
| ⑪ 1 | ⑫ 0 | | | |

| **38** | (p. 85 ~ 86) | | | |
|---|---|---|---|---|
| ① 10 | ② 41 | ③ 14 | ④ 17 | ⑤ 25 |
| ⑥ 13 | | | | |

$[(7 \times 10) - 32] \div 5 = 7R3$,

⑦ You put 7 cookies into each zip-loc bag, and 3 cookies were left.

⑧ $8 + (8 \times 5) + (8 + 9) = 65$, 65 apples

| **39** | (p. 87 ~ 88) | | | |
|---|---|---|---|---|
| ① 19 | ② 24 | ③ 0 | ④ 28 | ⑤ 83 |
| ⑥ 6 | ⑦ 23 | ⑧ 20 | ⑨ 10 | ⑩ 27 |

| **40** | (p. 89 ~ 90) | | | |
|---|---|---|---|---|
| ① 18 | ② 34 | ③ 70 | ④ 101 | ⑤ 54 |

$(37 \times 4) - (27 \times 5) = 13$

⑥ This week I ran 13 more minutes than last week.

$[99 - (7 \times 8) + 18] \div 6 = 10R1$

⑦ You put 10 marbles into each plate, and 1 marble was left.

# Tiger Math

# ACHIEVEMENT AWARD

## THIS AWARD IS PRESENTED TO

_____

(student name)

## FOR SUCESSFULLY COMPLETING

## TIGER MATH LEVEL F – 1.

*Dr. Tiger*

_____

Dr.Tiger